BAHAMAS PRIMARY

Mathematics Workbook 6

T0187389

The authors and publishers would like to thank the following members of the Teachers' Panel, who have assisted in the planning, content and development of the books:

Chairperson: Dr Joan Rolle, Senior Education Officer, Primary School Mathematics, Department of Education

Team members:

Lelani Burrows, Anglican Education Authority

Deidre Cooper, Catholic Board of Education

LeAnna T. Deveaux-Miller, T.G. Glover Professional Development and Research School

Dr Marcella Elliott Ferguson, University of The Bahamas

Theresa McPhee, Education Officer, High School Mathematics, Department of Education

Joelynn Stubbs, C.W. Sawyer Primary School

Dyontaleé Turnquest Rolle, Eva Hilton Primary School

Karen Morrison, Daphne Paizee and Rentia Pretorius

HODDER EDUCATION
AN HACHETTE UK COMPANY

Contents

Topic 1 Getting Ready

1 Measure each angle. Write the size on it in degrees. Complete the sentences.

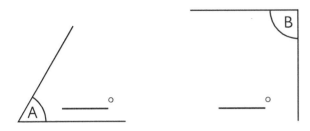 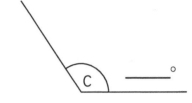

a Angle _____ is an obtuse angle.

b Angles which measure 90° are called _____ angles.

c Angle A is _____ because it measures _____ than 90°.

d Four right angles make a _____.

e A straight angle measures _____.

2 Each block on the grid is 1 cm long and 1 cm wide. On the grid, draw two different shapes both with an area of 9 square centimetres.

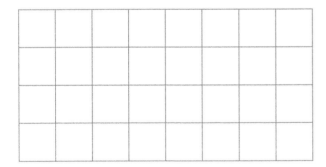

3 Toniqua did a survey of favourite colours among her friends.
Her results are:
red – 8,   blue – 2,   yellow – 5,
green – 6,   black – 7   and   purple – 3.
Complete the graph to show the information.

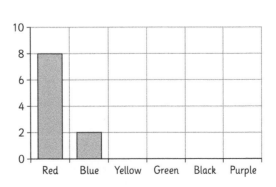

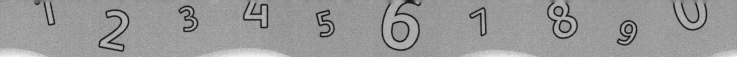

4 What must be added or subtracted to get the result shown? Fill in the operation and the amount.

    a 6.56 _____ _____ = 6.57

    b 6.56 _____ _____ = 6.46

    c 6.56 _____ _____ = 5.56

    d 6.56 _____ _____ = 6.01

    e 6.56 _____ _____ = 5.58

    f 6.56 _____ _____ = 6.023

5 Complete this addition pyramid. The number in each brick is found by adding the two below it.

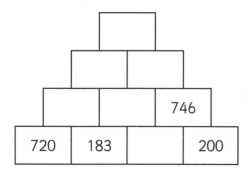

6 Draw lines on the diagrams to show how you could cut the octagon to make the shapes below it.

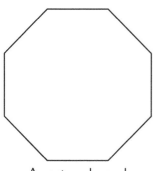
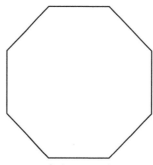

    A rectangle and   A kite and
    two trapeziums   two trapeziums

Topic 2 Numbers and Place Value

Place Value

1 Use the < or > symbols between each pair of numbers to compare then.

a 1 987 342 098 ☐ 1 879 342 098

b 12 324 098 312 ☐ 12 324 098 321

c 34 989 098 ☐ 34 989 098 123

d 19 099 865 432 ☐ 19 099 568 423

2 Write in numerals.

a forty-three million, two hundred fifty thousand, three hundred thirty-two.

_____

b nine billion, four hundred seven million.

_____

c forty-three billion, five hundred forty-seven million, eight hundred twelve thousand, six hundred forty-nine.

_____

d nine hundred ninety-nine billion, three hundred twenty-five million.

_____

3 Sort the numbers. Write them in the correct columns in the table.

123 987 098	13 298 709	9 999 999	12 098 098 456	123 456 876
32 432 000 000	432 000	10 098 765	1 000 876 543	9 000 000

Less than 10 million	Between 10 million and 1 billion	Greater than 1 billion

Working with Large Numbers

1 Use the figures in the table to complete the graph.

   a Give the graph a clear and suitable heading.

   b The point 3 billion is marked on the graph. Estimate the position of 6 billion and 9 billion and mark those points.

   c How old will you be when the population reaches the 9 billion mark?

   _____

Year	Population
1950	2 557 628 654
1960	3 043 001 508
1970	3 712 697 742
1980	4 444 496 764
1990	5 283 252 948
2000	6 008 279 216
2010	6 856 615 879
2020	7 629 798 111
2030	8 319 146 289
2040	8 897 252 335
2050	9 374 484 225

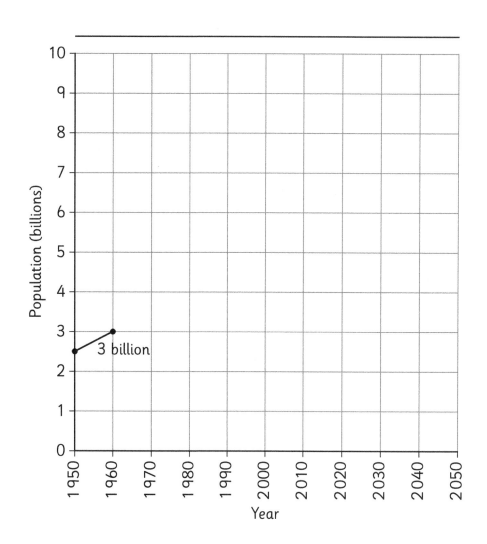

2 The world is divided into different regions. This graph shows how each region's share of the world population has changed over time and how it is likely to change in the future.

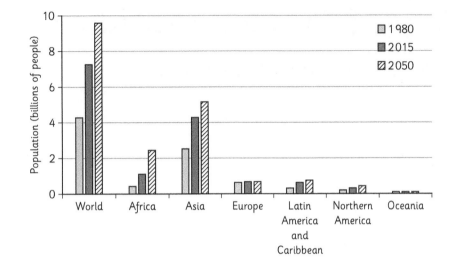

a Make up five questions that can be answered from the graph. Write your questions below.

b Exchange questions with a partner and try to answer each other's questions. Mark each other's answers.

Question 1 _____

_____

Answer _____

Question 2 _____

_____

Answer _____

Question 3 _____

_____

Answer _____

Question 4 _____

_____

Answer _____

Question 5 _____

_____

Answer _____

Topic 3 Exploring Patterns

Shape and Number Patterns

1 Look at the dot pattern.

   a Draw the next two shapes in this pattern.

   b Complete the table for the first five shapes.

Shape Number	1	2	3	4	5	n
Number of dots						

   c Look at the numbers in the bottom row of the table. What pattern can you see?

   _____

   d Write a rule in words for finding the number of dots in any shape (n).

   _____

   e How many dots would you need to draw the 10th shape? Write a number sentence
      with the answer.

   _____

2 Look at this pattern.

   a Draw the next shape.

   b Complete the table.

Shape Number	1	2	3	4	5	n
Number of dots						

   c Can you work out how many dots you would need to draw the 10th shape?
      And the nth shape? Complete the table.

Patterns in Calculations

1 Calculate the following and place the answers correctly in the table.

a

TTh	Th	H	T	O	.	t	h	th
				9	.	5	6	

9.56 × 10 = _____

9.56 × 100 = _____

9.56 ÷ 10 = _____

9.56 × 10 000 = _____

9.56 × 1000 = _____

b

TTh	Th	H	T	O	.	t	h	th
		1	2	3	.	4		

123.4 × 10 = _____

123.4 ÷ 100 = _____

123.4 ÷ 10 = _____

123.4 × 100 = _____

2 These calculations all result in 42.9. Fill in the missing values.

_____ ÷ 1000

_____ × 10

_____ ÷ 100 → 42.9 ← _____ ÷ 10

_____ × 1000

_____ × 100

Topic 4 Measurement

Using the Metric System

1 Complete the following chart.

Unit of Measurement	Abbreviation	Used to Measure ... (give an example)
Millimetre	mm	Length of a small insect
Metre		
Kilometre		
Milligram		
Gram		
Kilogram		
Millilitre		
Litre		
Kilolitre		

2 Measure these lines in cm and mm.

a _____

    i _____ cm

    ii _____ mm

b _____

    i _____ cm

    ii _____ mm

c _____

    i _____ cm

    ii _____ mm

d _____

    i _____ cm

    ii _____ mm

e _____

    i _____ cm

    ii _____ mm

f _____

    i _____ cm

    ii _____ mm

3 Write the following times using the 24 hour clock.

12 Hour Clock	24 Hour Clock
5:00 a.m.	
9:30 a.m.	
11:00 a.m.	
1:30 p.m.	

12 Hour Clock	24 Hour Clock
3:15 p.m.	
10:00 p.m.	
11:20 p.m.	
12:05 a.m.	

4 Measure the temperature in your classroom every morning when you arrive at school and then just before you leave school every afternoon. Record your measurements on the following chart. Make sure you write the unit of measurement that you use (°C or °F).

Days	At _____ a.m.	At _____ p.m.	Difference
Mon.			
Tues.			
Wed.			
Thurs.			
Fri.			

5 Draw a graph to show the data you have collected.

Converting Units

1 Convert the following measurements:

a 3.5 kg = _____ g

b 250 g = _____ kg

c _____ g = 0.3 kg

d 0.05 g = _____ kg

e _____ mL = 12 L

f 5 600 mL = _____ L

g 0.125 L = _____ mL

h 2.5 L = _____ mL

i 0.5 km = _____ m

j 7 500 m = _____ km

k 65 mm = _____ cm

l _____ cm = 4.7 m

2 Look at the quantities given in this recipe. Use the conversion table to answer the questions. You can use a calculator to help you.

Ingredients for Pancakes
4 oz flour
10 fl oz milk
2 oz butter
a pinch of salt

Conversion Table

Customary Units	Metric Units
1 ounce (oz)	28.35 grams
1 fluid ounce (fl oz)	29.57 millilitres

Underline the best answer.

a How much flour will you need?
Approximately:
i 113 grams   **ii** 400 grams   **iii** 313 grams

b How much milk will you need?
Approximately:
i 300 mL   **ii** 30 L   **iii** 30 mL

c Will you need:
i more than 100 g of butter   **ii** less than 100 g of butter?

d How much is a 'pinch of salt'?
i less than one gram   **ii** more than one gram   **iii** about one kg

Apply Your Skills

1 Work with a partner. For each quantity in the table, write down:

    a what instrument you would use to measure it

    b what units you would measure it in

    c an estimated measurement

    d an actual measurement (you will need to measure the quantities to do this).

Quantity	Instrument I Would Use	Units I Would Measure In	My Estimate	My Measurement
The width of the classroom				
The height of my teacher				
The length of my pen				
The distance from the door to the gate				
The mass of a book				
The mass of a calculator				
The mass of a litre of water				
The amount of water a bucket can hold				
The amount of coffee in a mug				
How long it takes to count to ten				
How long it takes me to sign my name				
How long it takes to walk 100 metres				
The temperature in the classroom				
The temperature outside in the shade				

Topic 5 Fractions

Equivalent Fractions

1 This grid shows a fraction wall but the labels have fallen off. Label each part to show what fraction it represents.

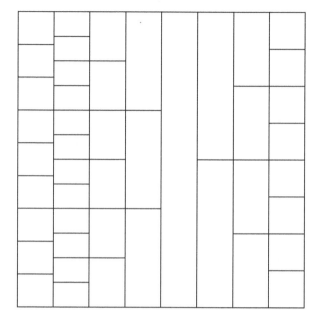

2 Use the grid to find an equivalent fraction in simplest terms for each of these fractions.

a $\frac{2}{6} = \frac{\square}{\square}$

b $\frac{4}{6} = \frac{\square}{\square}$

c $\frac{6}{9} = \frac{\square}{\square}$

d $\frac{8}{12} = \frac{\square}{\square}$

e $\frac{9}{12} = \frac{\square}{\square}$

f $\frac{6}{8} = \frac{\square}{\square}$

g $\frac{5}{10} = \frac{\square}{\square}$

h $\frac{3}{12} = \frac{\square}{\square}$

3 Use the wall to compare these fractions. Fill in < or >.

a $\frac{1}{2} \,\square\, \frac{1}{3}$

b $\frac{2}{3} \,\square\, \frac{6}{10}$

c $\frac{5}{12} \,\square\, \frac{4}{10}$

d $\frac{3}{4} \,\square\, \frac{8}{10}$

e $\frac{1}{5} \,\square\, \frac{2}{12}$

f $\frac{7}{10} \,\square\, \frac{2}{3}$

g $\frac{1}{4} \,\square\, \frac{5}{12}$

h $\frac{3}{4} \,\square\, \frac{2}{3}$

i $\frac{9}{12} \,\square\, \frac{4}{5}$

j $\frac{3}{5} \,\square\, \frac{2}{3}$

Investigate Fractions

- Colour in a section of each grid. All your grids must be different.
- Write the fraction of the grid that is shaded with a denominator of 24.
- Write any simpler, equivalent fractions that you can.
- The first one has been done as an example.

a

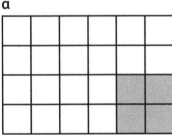

$$\frac{4}{24} = \frac{2}{12} = \frac{1}{6}$$

b

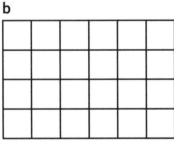

c

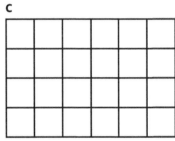

d

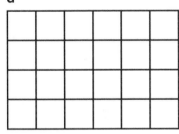

e

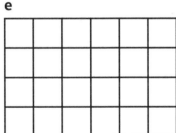

f

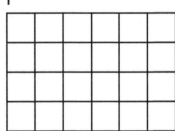

g

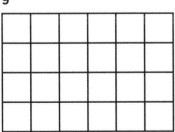

h

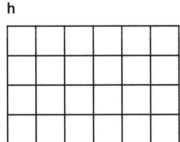

i

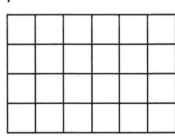

Topic 6 Integers

Positive and Negative Integers

1 Plot the integers on the number line.

a 6, −3, 8, 3, −9

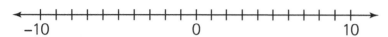

b 20, −10, 5, 0, −25

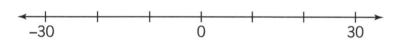

2 Order each set of numbers from the least to the greatest.

a 1, −8, 7 _____

b −2, 20, 17, −13 _____

c 4, −6, 10, 6, 0, −1 _____

3 Draw the following items at the correct heights on the diagram.

a A ball 2 m above the water level.

b A conch on a rock 2 m below the water level.

c A seagull flying 3 m above the conch.

d A fish $1\frac{1}{2}$ m below the water level.

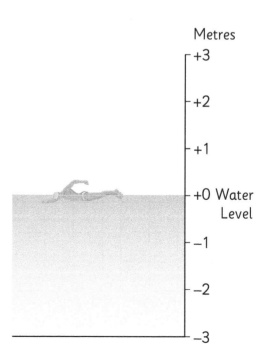

Topic 7 Decimals and Percentages

Revisit Decimals

1 Use the key below to colour the digit in each place value for each number.

thousands	red	tenths	blue
hundreds	orange	hundredths	purple
tens	yellow	thousandths	pink
units	green		

a 7 2 4 1 . 3 0 2   **b** 1 5 . 0 0 5

c 2 0 2 0 . 8   **d** 2 3 7 . 7 1

e 6 . 2   **f** 2 0 0 9 . 0 0 8

2 Use the grids provided to represent the number by shading the correct number of blocks and then write the decimal equivalent below each diagram.

a $\frac{5}{10}$

_____

b $\frac{10}{10}$

_____

c $\frac{9}{100}$

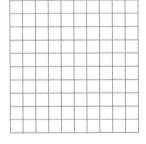

_____

d $\frac{47}{100}$

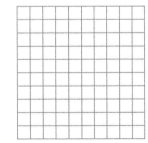

_____

Percentages

1 Shade the correct number of squares to show each percentage.

a 55%

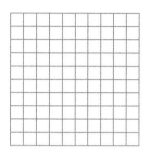

b 1%

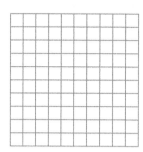

c 92%

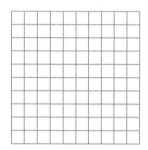

d 114%

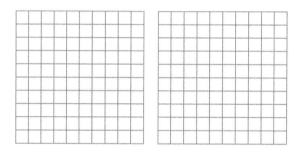

2 Complete the table below by inserting equivalent terms.

Percent	Fraction	Decimal
6%		
		0.44
	$\frac{19}{20}$	
		0.013
155%		

Topic 8 Classifying Shapes

Angles

1 Write down the names and the measurements of each angle.

a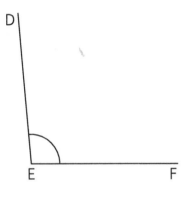

b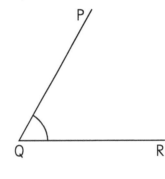

∠ _____ = _____ °

∠ _____ = _____ °

c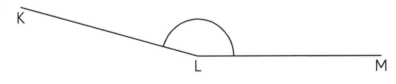

∠ _____ = _____ °

2 Draw an example of each type of angle.

a Obtuse angle	**b** Straight angle
c Right angle	**d** Acute angle

Triangles

1 Draw the correct markings on each type of triangle to show its properties.

Scalene triangle	Isosceles triangle
Equilateral triangle	Right-angled triangle

2 Draw a scalene right-angled triangle. Use only a ruler and a pencil.

3 Write TRUE or FALSE next to each statement.

a A triangle can be equilateral and scalene. _____

b A triangle can be right-angled and obtuse angled. _____

c A triangle can be isosceles and right-angled. _____

d Half of a square will always be a right-angled triangle. _____

Quadrilaterals

1 Complete the table by ticking all properties that apply to each shape.

Property	Square	Rectangle	Parallelogram	Rhombus	Trapezium
All sides are equal					
Opposite sides are equal					
All sides are parallel					
Opposite sides are parallel					
One pair of sides is parallel					
All angles are equal					
Opposite angles are equal					

2 Investigation: Is a square a parallelogram?

   a Is a square a rectangle? Write down the definition of a rectangle.

   _____

   b Draw a square on the right, labelling equal
   and parallel sides as well as the
   right angles.

   c Can you say that a square is a special type of rectangle?
   Circle   YES   or   NO

   d Is a rectangle a parallelogram? Write down the definition of a parallelogram.

   _____

e Draw a rectangle on the right, labelling
equal and parallel sides, as well as
the right angles.

f Does a rectangle match the characteristics
of a parallelogram?
Circle YES or NO

g Is a square a parallelogram? What is your conclusion?

_____

h Now state whether the following statements are TRUE or FALSE.

- All rectangles are parallelograms.   _____

- All parallelograms are rectangles.   _____

- All squares are rectangles.   _____

- All rectangles are squares.   _____

3 You can make shapes on a pegboard using elastic bands. Use colours to show three
different ways of making each shape on a pegboard. One shape has been drawn for you.

a Rectangle

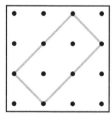

b Parallelogram

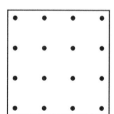

c Rhombus

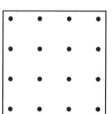

4 One way of making a square is shown here. How many different ways can you find
to make a square? Draw them using different colours.

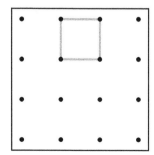
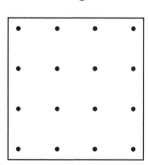
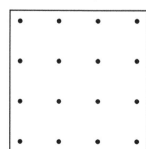

Circles

1 Identify the circle parts as shown below.

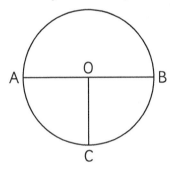

a O is the

_____.

b AB is the

_____.

c OC is the

_____.

d OB is the

_____.

2 Use different colours to show the parts of the circle.

a Centre and radius

b Diameter

c Circumference   _____   _____   _____

Investigate Shapes

1 Complete this table by filling in the number of sides that each polygon has and then writing where you can see an example of a regular and irregular version of each one in the real world.

Number of Sides	Name of Polygon	A Regular Example from the Real World	An Irregular Example from the Real World
	Triangle		
	Quadrilateral		
	Pentagon		
	Hexagon		
	Heptagon		
	Octagon		
	Nonagon		
	Decagon		

2 Find an example of something in real life that matches each shape.

A Small Sphere	A Short Fat Cylinder	A Tall Thin Cylinder
A Pyramid	A Large Cuboid	A Small Cube

Topic 9 Rounding and Estimating

Round Numbers

1 Round each number to the place given in the table.

Number	Rounded to the nearest ...					
	Ten	Thousand	Million	Hundred	Hundred Thousand	Ten Million
12 345 678						
19 099 087						
1 234 987 098						
29 098 765						
146 098 554						

2 Round each decimal to the place given in the table.

Decimal	Hundredth	Whole Number	Tenth	Ten
29.876				
13.098				
19.992				
25.044				
27.187				
9.999				

3 Use the digits on the cards and the decimal point (if you need it) to make numbers that will round to the numbers a–e. Use the digits once only in each number you make.

3	7	2	6	·

a _____ will give 3.6 when rounded to the nearest tenth.

b _____ will give 2 when rounded to the nearest whole number.

c _____ will give 2 000 when rounded to the nearest thousand.

d _____ will give 7.4 when rounded to the nearest tenth.

e _____ will give 6.33 when rounded to the nearest hundredth.

Estimating

For each receipt:

    a Use the method you find most efficient to round the amounts and then estimate the total of each bill in whole dollars.

    b Estimate each person's equal share of the total for the given number of people.

Use the space next to each receipt to show your working.

1

Jo's Conch

$17.99

$54.20

$25.00

$15.50

$84.75

Four people.

Each owes approximately _____

2

Mae's Munchies

$ 9.99

$ 4.50

$12.99

$ 5.80

$ 7.50

$ 9.25

Two people.

Each owes approximately _____

3

Sanita's Sweets

$3.80

$4.26

$9.50

$6.99

$2.99

$3.50

$4.20

$8.00

Five people.

Each owes approximately _____

4

The Poop Deck

$ 35.80

$ 73.96

$ 29.99

$ 42.80

$ 63.75

$122.80

Six people.

Each owes approximately _____

Topic 10 Mental Methods

Complete each grid as quickly as you can. Pay attention to the operation signs.

a

×	2	5	7	8	10	6	4	3	9
6									
4									
2									
7									
8									
3									
9									
11									
12									
10									
20									

b

	11	28	45	90	99	12	88	100	55
100 –									
150 –									
200 –									
200 +									
55 +									
120 +									

c Divide the number in the top row by the number in the first column.

÷	30	300	60	600	90	120	450	4500	1800
2									
10									
5									
3									
6									

Time Trials

1 Time yourself to see how quickly you can complete each set of facts.
  Write the answers only. Record your time and how many you got correct.

2 Cover your answers for the first set and try again. Compare your results.

Calculate	First Round Answers	Second Round Answers
36 ÷ 6		
36 ÷ 12		
100 ÷ 5		
25 ÷ 25		
15 ÷ 3		
42 ÷ 7		
72 ÷ 8		
63 ÷ 7		
81 ÷ 9		
63 ÷ 9		
Time:		
Number correct:		

Calculate	First Round Answers	Second Round Answers
32 × 10		
129 × 10		
14 × 5		
12 × 50		
13 × 100		
120 × 3		
19 × 10		
19 × 5		
30 × 12		
20 × 42		
Time:		
Number correct:		

Calculate	First Round Answers	Second Round Answers
53 + 27		
19 + 87		
12 + 88		
17 + 42		
120 + 180		
130 + 125		
200 + 87		
125 + 95		
400 + 230		
88 + 102		
Time:		
Number correct:		

Calculate	First Round Answers	Second Round Answers
100 − 55		
60 − 24		
120 − 90		
130 − 88		
100 − 72		
400 − 32		
300 − 185		
120 − 60		
300 − 149		
200 − 99		
Time:		
Number correct:		

Number Pyramids

In each number pyramid, a block is the sum of the two blocks below it.

Work out the missing values to complete each pyramid. Use the space below the pyramid for working, if you need it.

a

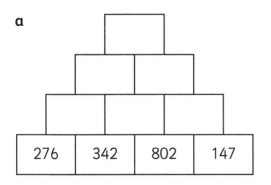

b

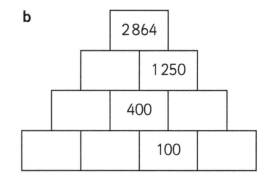

c

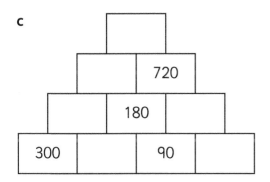

d

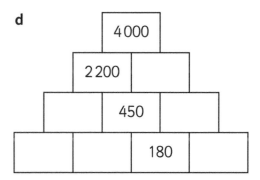

Topic 11 Factors and Multiples

Factors

1 Write the factor pairs for each number.

a
10	

b
18	

c
24	

2 Check these lists of factors. Cross out any incorrect numbers. Write in any missing factors.

a

Factors of 40
1 2   4   5   6
10   20   40

b

Factors of 48
1   2   4   6
12   48

c

Factors of 64
1   2   4   8   16
34   64

d

Factors of 72
1   2   3   4   6   8   9
12   36   72

e

Factors of 81
1   3   9   80

f

Factors of 96
1   2   3   4   5   6
12   16   24   48   96

Multiples

1 Write the missing multiples in each set.

  a 4, 8, 16, _____, _____, _____, _____, 36, _____

  b 16, 20, _____, _____, 32, 36, _____, _____

  c _____, 18, 27, _____, _____, _____, _____, 72

  d _____, 24, _____, 40, _____, _____, _____, 72

2 Write the multiples.

  a 5th multiple of 8 _____   **b** 6th multiple of 9 _____

  c 12th multiple of 3 _____   **d** 100th multiple of 7 _____

3 Tick the columns that apply to each number.

	Multiple of 2	Multiple of 3	Multiple of 4	Multiple of 5	Multiple of 2 and 3	Multiple of 3 and 4	Multiple of 4 and 5
24							
9							
18							
30							
36							
50							
11							
48							
16							
25							
20							
22							
60							

GCF and LCM

1 Complete the factor trees and write each number as the product of its prime factors.

a

12 = _____

b
16

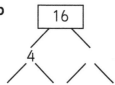

16 = _____

c
18

18 = _____

d
20

20 = _____

e
24

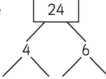

24 = _____

f
252

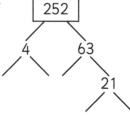

252 = _____

2 Use the information in question 1 to find the GCF of:

a 12 and 20 _____

b 16 and 20 _____

c 24 and 252 _____

d 252 and 18 _____

3 What is the LCM of:

a 12 and 20 _____

b 12 and 24 _____

c 16 and 24 _____

d 24 and 252 _____

Topic 12 Scale and Distance

1 Complete the statements.

   a If the scale is 1 : 100, then 1 cm on the map = _____ in reality.

   b If the scale is 1 : 1 000, then 1 cm on the map = _____ in reality.

   c If the scale is 1 : 1 000, then 1 mm on the map = _____ in reality.

   d If the scale is 1 : 10 000, then 1 cm on the map = _____ in reality.

2 Mario is going to do a scale drawing using a scale of 1 : 100.
Draw a line to show accurately what length each of these real distances would be on his diagram.

   a 100 cm

   b 750 cm

   c 4.5 m

   d 15 m

   e 10.2 m

3 Measure the length of each line between each pair of points. Complete the table.

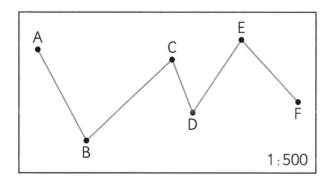

1 : 500

	A to B	B to C	C to D	D to E	E to F
Distance on Diagram					
Distance in Reality					

Working with Scale

1 The diagram below shows the route followed by a taxi driver each weekday.

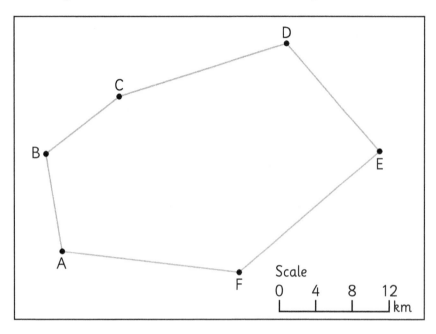

Measure the distances on the map and complete this table for his route.

Route	Distance on Map	Distance in Real Life
A to B		
B to C		
C to D		
D to E		
E to F		
F to A		
Whole route		

2 If the driver does this route eight times a day, what distance (in kilometres) does he cover in real life? Show how you worked this out.

_____

_____

Make a Scale Drawing

Pick one room in your house. You will draw a map to scale of the room on the 1 cm grid.

Measure the room and the larger items in the room. Once you have made your measurements, decide on a scale and work out what size you should draw the walls and the items on the grid. **Remember to include your scale on the map.**

Topic 13 Graphs

Organizing and Representing Data

1 Draw a circle graph using the following data. You will need to work out the angles and use your protractor to draw these angles. Provide a key, or label each sector of the circle graph.

The Brand of Chocolate that Customers Bought in the Last Month

Brand	Number	Angle
Oh Henry	30	
Mars	6	
60% Pure	10	
Bates Bar	14	

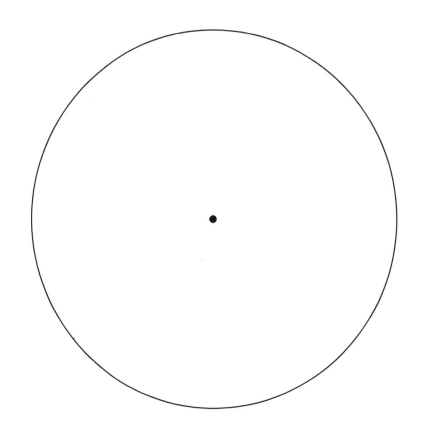

2 Draw a double bar graph with the given Olympic medal data below.

2012 Olympic Games

Country	Gold Medals
United States	46
China	38
Great Britain	29
Germany	11
Australia	7
Bahamas	1

2016 Olympic Games

Country	Gold Medals
United States	46
Great Britain	27
China	26
Germany	17
Australia	8
Bahamas	1

3 Estimate the monthly and average rainfall for each month as accurately as you can and then draw a **line graph** of the data shown in the bar graph below.

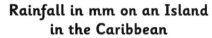

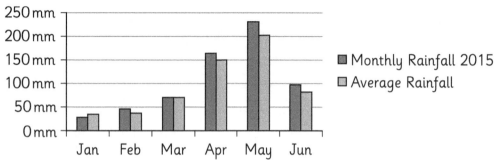

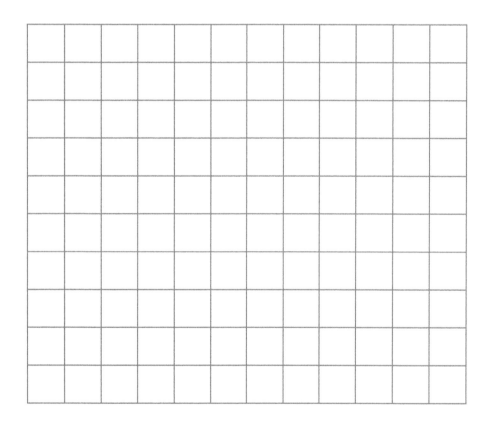

4 Use this page to draw graphs to represent the data from question 6 in your Student Book.

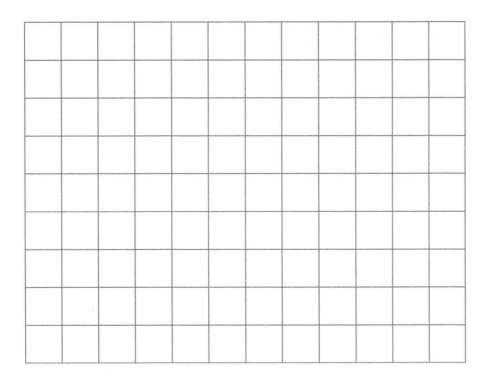

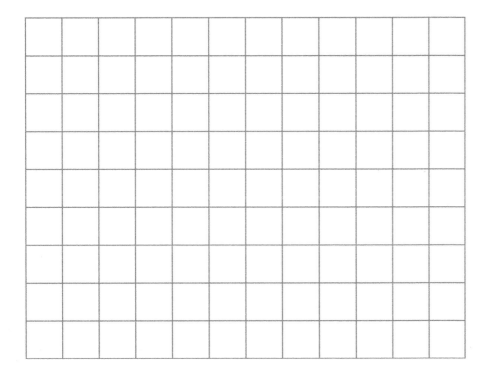

Analyse Data

The table and graph give you some information about the number of medals won by five countries in the Summer Olympics from 2000 to 2016.

Use the information to answer the questions.

Country	Total Medals Won				
	2000	2004	2008	2012	2016
USA	93	101	110	103	121
China	58	63	101	88	70
Great Britain	28	30	47	65	67
Russia	89	90	62	72	55
Germany	56	49	41	44	42

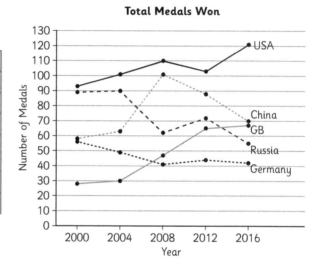

Total Medals Won

a What is the range of the number of medals collected by the top 5 countries in 2016?

_____

b What is the mean number of medals collected by the United States and by China at the five different Olympic Games competitions?

United States: _____

China: _____

c What is the median of the number of medals collected by China in the five different games?

_____

d What trends does the graph show?

_____

Topic 14 Adding and Subtracting

1 Add.

a   377
   +423
   ————

   ————

b   489
   +575
   ————

   ————

c   789
   +204
   ————

   ————

d   1 204
  + 987
  ————

  ————

e   19 246
  +11 432
  —————

  —————

f   14 089
  + 9 264
  —————

  —————

2 Subtract.

a   486
  −239
  ————

  ————

b   499
  −207
  ————

  ————

c   784
  −396
  ————

  ————

d   2 074
  −1 042
  ————

  ————

e   2 000
  −1 486
  ————

  ————

f   12 072
  − 9 387
  —————

  —————

3 What is:

a $5\,000 + 264 =$ ————————

b $10\,000 - 2\,500 =$ ————————

c $70\,000 + 40\,000 =$ ————————

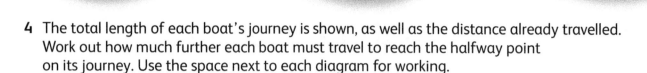

4 The total length of each boat's journey is shown, as well as the distance already travelled. Work out how much further each boat must travel to reach the halfway point on its journey. Use the space next to each diagram for working.

a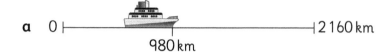

0 ⊢———————————————⊣ 2160 km
      980 km

Answer: _____

b

0 ⊢———————————————⊣ 4480 km
      1882 km

Answer: _____

c

0 ⊢———————————————⊣ 1445 km
      487 km

Answer: _____

d

0 ⊢———————————————⊣ 2812 km
      742 km

Answer: _____

e

0 ⊢———————————————⊣ 4875 km
      1799 km

Answer: _____

f

0 ⊢———————————————⊣ 3638 km
      1689 km

Answer: _____

Topic 15 Problem Solving

Number Puzzles

In this number puzzle, the number in each oval is the sum of the two numbers in the circles on either side of it. Can you find six different solutions that work?

a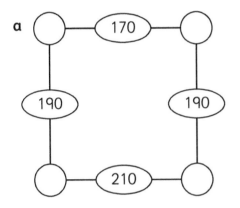

b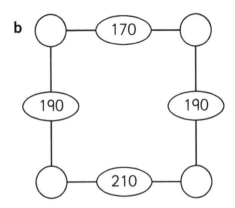

c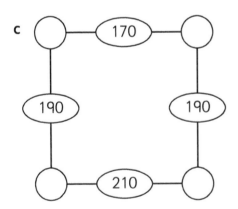

d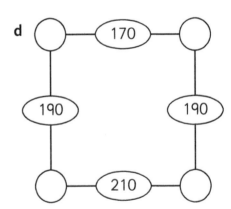

e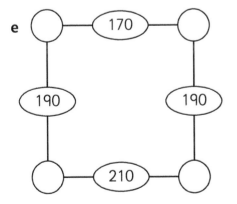

f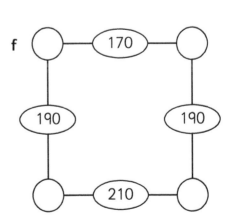

Find the Way

In this number puzzle, you may move along any arrow and add or subtract the number in the bubble as you go.

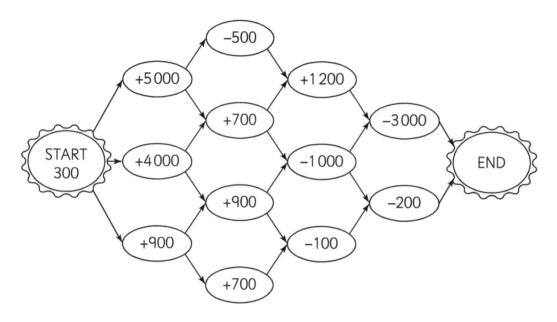

Find routes through the puzzle that give the following totals.

Write the route you followed.

The route for 1 000 is given as an example.

Route for 1 000: $\underline{300 + 4\,000 + 700 - 1\,000 - 3\,000 = 1\,000}$

a Route for 2 000:   _____

b Route for 3 000:   _____

c Route for 4 000:   _____

d Route with highest possible total: _____

e Route with lowest possible total: _____

Expressions and Equations

1 Find the value of x and y in each flow diagram.

Write an equation for each calculation. The first one has been done for you.

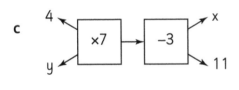

a

x → +5 → ×2 → 24
6 ← ← ← y

$$(24 \div 2) - 5 = 7$$

$$(6 + 5) \times 2 = 22$$

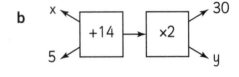

b

x → +14 → ×2 → 30
5 ← ← ← y

_____

_____

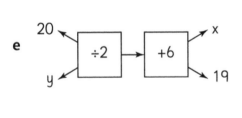

c

4 → ×7 → −3 → x
y ← ← ← 11

_____

_____

d

9 → ×4 → −4 → x
y ← ← ← 8

_____

_____

e

20 → ÷2 → +6 → x
y ← ← ← 19

_____

_____

f

x → ×3 → +5 → 20
6 ← ← ← y

_____

_____

2 Write all the pairs of values of x and y that make this equation true.

$9 + 9 = x \times y$

_____

_____

More Problems

1 Follow the flow diagram instructions.

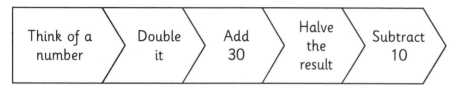

Try this with 5 different starting numbers.

What do you notice?

_____

_____

Try to explain the results.

_____

_____

2 Follow the flow diagram instructions.

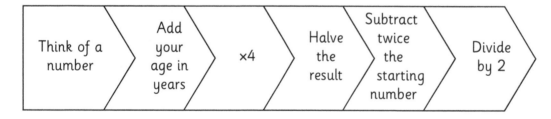

Try this with 5 different starting numbers.

What do you notice?

_____

_____

Try to explain the results.

_____

_____

Topic 16 Ratio

Describing Ratios

1 Colour in the following pictures using the given ratios of blue to red sections.

a 3:5   **b** 2:6   **c** 1:4

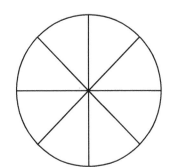 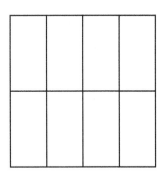 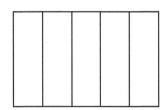

2 Complete the table by writing each ratio in their simplest (equivalent) forms.

Ratio	Equivalent Form
3:9	
30:60	
15 mm:45 mm	
0.25:0.50	
72:36	
66:121	
40%:60%	
0.10:0.100	

3 Complete the sentences.

a The ratio of concentrated juice to water in a jug is 1:8.

For every one glass of concentrated juice, I added _____ glasses of water.

Now I have enough juice to fill _____ glasses.

b Seymour spent five hours running and walking yesterday.

The ratio of his running and walking was 4:1, so he spent _____

hours running and _____ hours walking.

Pi

1 Work in pairs. Find six different objects with circles; for example: cups, bowls, clocks, pizzas. Measure the circumference of each circle. Measure the diameter. Then calculate the ratio (pi). Record your measurements and calculations in the table below.

Item	Circumference	Diameter	Pi

2 Draw the following circles. Use the given measurements. Then complete the labels for each circle.

a Radius: 3.5 cm

Diameter: _____

Circumference: _____

b Radius: _____

Diameter: 40 mm

Circumference: _____

c Radius: _____

Diameter: _____

Circumference: 15.7 cm

Scale

1 Use the graph paper below to make four scale drawings of objects in your classroom.

Write the scale under each drawing.

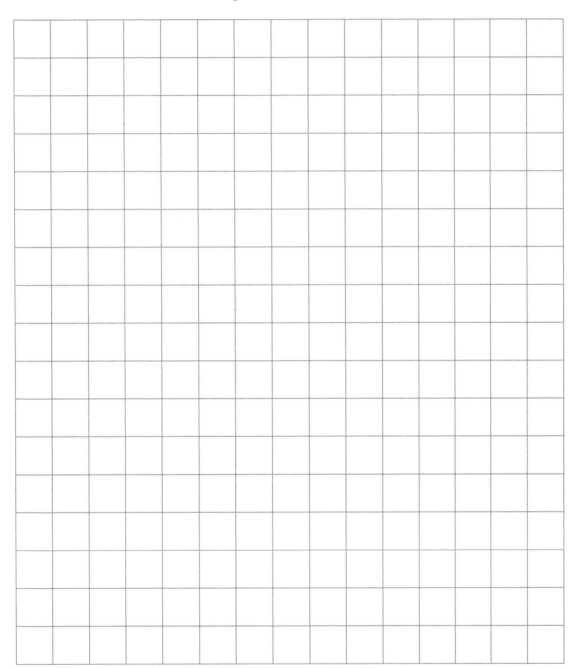

2 Look at the following two maps, both of which show Eleuthera Island.

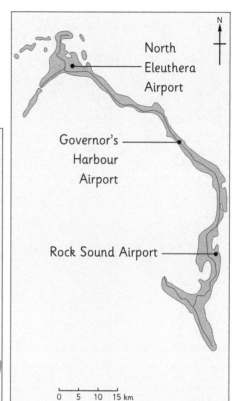

Islands of The Commonwealth of The Bahamas

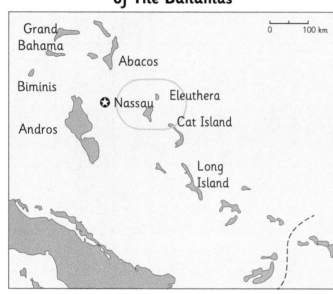

Scale: 1 cm = _____

Scale: 1 cm = _____

a Use the scale bar to work out and complete the scale below each map.

b Why were different scales used for these maps?

_____

_____

c Explain in your own words why scale is a ratio.

_____

_____

Topic 17 Multiplying and Dividing

Mental Methods

1 Work out and fill in **all** the missing numbers in these multiplication squares.

✕	5		
4		28	
		35	
9			81

✕	3		
6			
	21		
9		54	72

✕	8		
4			
	24	18	
8			80

✕	7		
2			18
	42		
10		80	

2 Estimate the products and write them in descending order. You do not need to do the actual calculations.

| 8 × 77 | 4 × 155 | 9 × 66 | 3 × 199 |

| 5 × 128 | 7 × 81 | 6 × 99 | 3 × 174 |

_____, _____, _____, _____, _____, _____, _____, _____

3 Find the quotient. Write two related multiplication facts for each one.

a 140 ÷ 20 = _____   _____   _____

b 60 ÷ 3 =   _____   _____   _____

c 140 ÷ 5 =   _____   _____   _____

d 128 ÷ 4 =   _____   _____   _____

e 450 ÷ 6 =   _____   _____   _____

f 350 ÷ 7 =   _____   _____   _____

g 85 ÷ 5 =   _____   _____   _____

4 In these triangles, the top number is the product of the bottom two. Work out the missing numbers.

a
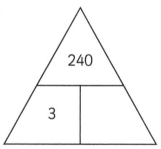

b
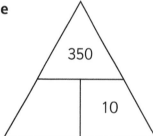

c
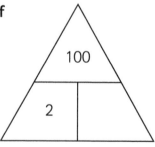

d
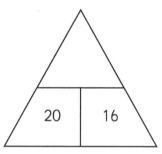

e
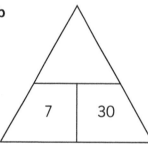

f
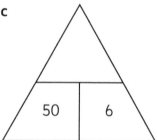

g
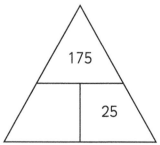

h
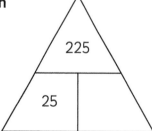

i
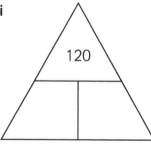

Multiplying Larger Numbers

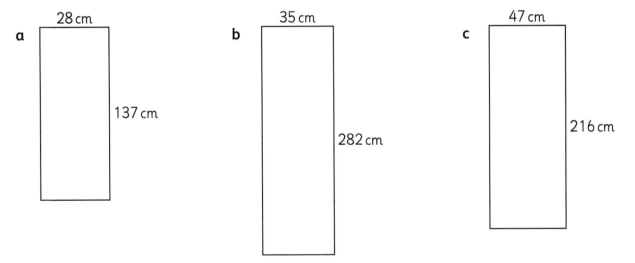

a   28 cm   137 cm

b   35 cm   282 cm

c   47 cm   216 cm

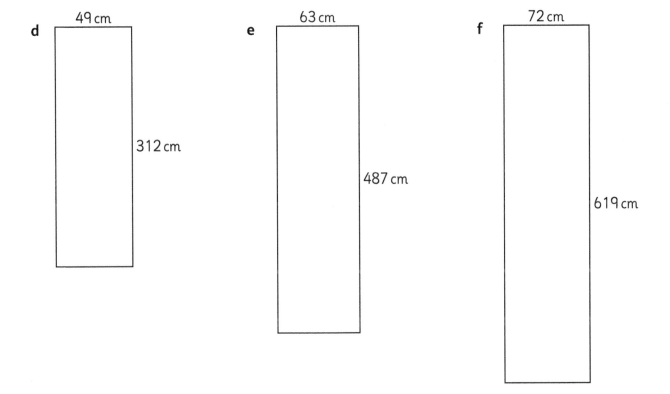

d   49 cm   312 cm

e   63 cm   487 cm

f   72 cm   619 cm

Work out the area of each rectangle. Show your working out inside each rectangle.

Dividing Larger Numbers

Complete each division.

a 1 5) 7 4 3   **b** 1 6) 6 4 8   **c** 2 8) 8 6 1

d 1 3) 3 3 3   **e** 1 2) 4 1 6   **f** 2 4) 6 8 7

g 4 2) 4 3 1   **h** 2 2) 7 3 2   **i** 3 5) 5 1 7

j 3 1) 6 3 4   **k** 4 5) 9 9 9   **l** 2 4) 1 0 8 7

Topic 18 Exploring Shape

Lines of Symmetry

1 For each shape below, does the dotted line show a line of symmetry?
Write yes or no.

a 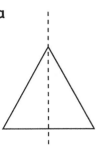 **b** 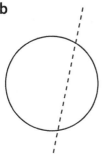 **c**  **d** 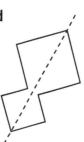 **e**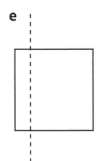

2 Draw in **all** the lines of symmetry on each of these shapes.

a  **b** 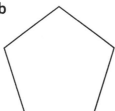 **c**

d 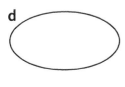 **e** 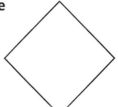 **f**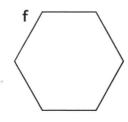

g  **h** 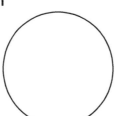 **i**

3 Draw in the other half of each symmetrical shape below.

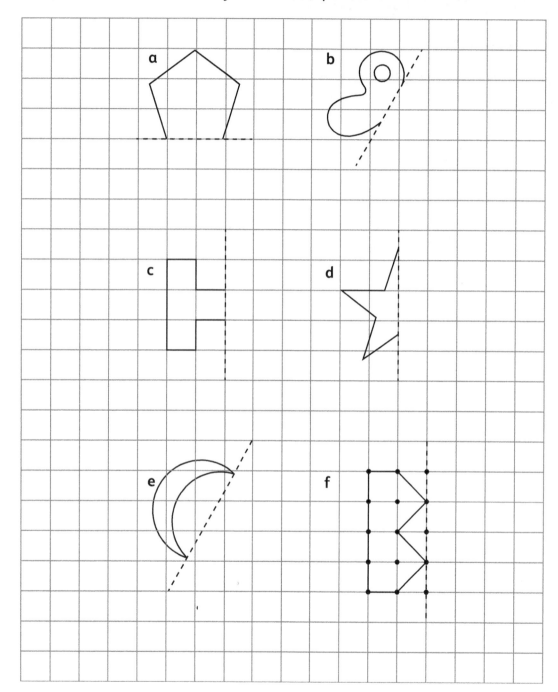

a

b

c

d

e

f

4 Investigate the shapes below. For each shape, write the name of the shape, the number of sides and how many lines of symmetry it has.

a

Name of shape: _____

Number of sides: _____

Number of lines of symmetry: _____

b

Name of shape: _____

Number of sides: _____

Number of lines of symmetry: _____

c

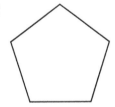

Name of shape: _____

Number of sides: _____

Number of lines of symmetry: _____

d

Name of shape: _____

Number of sides: _____

Number of lines of symmetry: _____

e

Name of shape: _____

Number of sides: _____

Number of lines of symmetry: _____

f

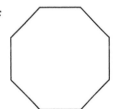

Name of shape: _____

Number of sides: _____

Number of lines of symmetry: _____

Translation

1 Another word for a translation is a _____.

2 Translate each polygon according to the instructions below.

- Shape ABC   translate 4 units up.

- Shape DEFG   translate 2 units down.

- Shape HIJ   translate 1 unit down and five units to the right.

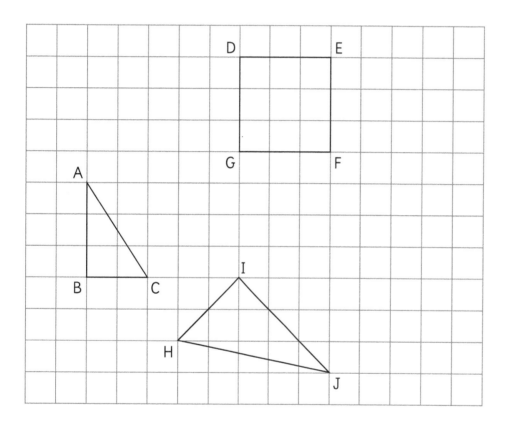

3 Create a pattern in this block using translation of two different shapes.

Reflection

1 Another word for a reflection is a _____.

When you reflect a shape, each point is reflected in the _____ line.

2 Complete each reflection.

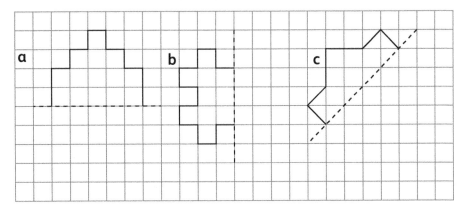

3 Now draw your own reflections of these shapes.

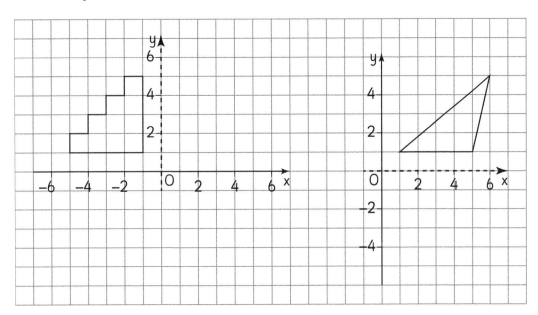

4 Explain in your own words the relationship between reflections and line symmetry.

_____

_____

Rotation

1 Another word for a rotation is a _____ .

When you rotate a shape, you can describe three things:

_____

_____

_____

2 Rotate this shape 90° about point K. Repeat the rotation until it gets back to the original object.

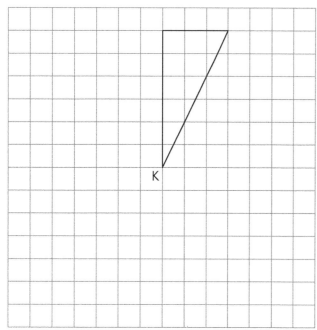

3 Create your own pattern by rotating a shape about the point P.

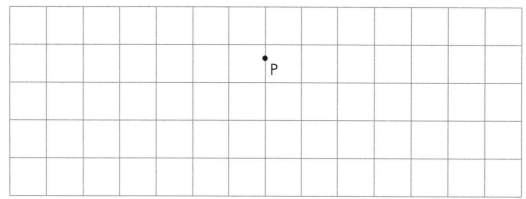

Topic 19 Order of Operations

What Do You Do First?

For each calculation, circle the part that you would work out first. Then calculate the answers.

a   20 + 5 − 5 × 5 =

b   8 ÷ 2 + 6 =

c   3 + 6 − 4 × 2 =

d   56 + 4 ÷ 2 =

e   12 − 2 + 22 ÷ 2 − 1 =

f   44 ÷ (6 − 2) + 4 × 2 + 5 =

g   3 × 10 + 12 − 4 =

h   56 ÷ 8 + 3 × 4 =

i   (27 + 9) ÷ (12 − 9) =

j   40 + 20 × 3 =

k   3 × (5 + 2 + 4) ÷ 3 =

l   (3 + 8) × 2 + 6 =

m   10 × 10 − 40 + 10 =

Time Trials

For each set of calculations, work out the answers as quickly as you can. Record your time and score. Remember to work in the correct order.

Set A	Set B
(12 – 5) × 3 =	6 + 8 × 2 =
30 ÷ 3 + 3 =	6 × 3 – 12 =
8 + 3 × 7 =	3 × (4 + 6) =
6 – 5 × 0 =	(9 – 5) × 8 =
4 × 8 – 4 =	4 × 6 – 7 =
5 × 3 – 2 =	7 + 48 ÷ 4 =
(8 – 2) × 5 =	14 + 12 ÷ 3 =
24 ÷ 3 – 5 =	7 × (12 – 7) =
2 × 4 × 3 =	18 ÷ 3 × 4 =
12 ÷ 2 × 3 =	(17 – 9) × 7 =
12 – 5 × 2 =	4 × (3 + 5) =
9 + 12 ÷ 2 =	12 × 3 + 11 =
4 + 15 ÷ 3 =	48 ÷ (2 × 4) =
5 × 4 ÷ 2 =	28 – 4 × 3 =
14 – 6 × 2 =	12 + 4 – 8 × 0 =
Time taken: Score: $\frac{}{15}$	Time taken: Score: $\frac{}{15}$

Grouping Symbols

1 The answers to these calculations are correct. However, the brackets have been left out of some of them. Check the calculations and insert brackets if they are needed.

a 4 + 8 × 10 = 120

b 19 − 9 ÷ 5 = 2

c 25 − 50 − 30 = 5

d 8 × 4 + 5 = 72

e 4 + 2 × 7 = 42

f 20 − 14 × 8 = 48

g 14 + 36 ÷ 6 = 20

h 16 + 12 ÷ 4 = 7

i 70 ÷ 35 ÷ 5 = 10

j 93 − 12 ÷ 3 = 27

k 40 ÷ 13 − 8 = 8

l 40 ÷ 5 × 4 = 2

m 15 + 100 ÷ 20 = 20

n 16 − 9 + 23 = 30

o 15 + 100 ÷ 5 = 35

p 100 − 7 × 5 = 65

q 20 − 28 − 19 = 11

r 45 − 64 ÷ 8 = 37

s 10 ÷ 28 − 23 = 2

t 6 + 2 × 8 + 2 = 80

Topic 20 Calculating with Fractions

Add and Subtract Fractions and Mixed Numbers

1 The distances cycled by four cyclists are shown.

Write a calculation and work out the total distance each one cycled.

a

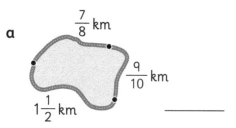

$\frac{7}{8}$ km

$\frac{9}{10}$ km

$1\frac{1}{2}$ km _____

b

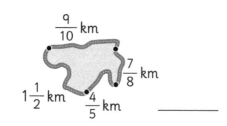

$\frac{9}{10}$ km

$\frac{7}{8}$ km

$1\frac{1}{2}$ km $\frac{4}{5}$ km _____

c

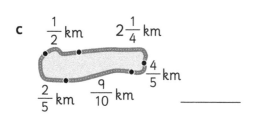

$\frac{1}{2}$ km $2\frac{1}{4}$ km

$\frac{4}{5}$ km

$\frac{2}{5}$ km $\frac{9}{10}$ km _____

d

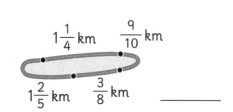

$1\frac{1}{4}$ km $\frac{9}{10}$ km

$1\frac{2}{5}$ km $\frac{3}{8}$ km _____

2 The total distance and the distance already cycled for four different routes are shown here.

Work out how much further each cyclist has to go to complete the route.

a

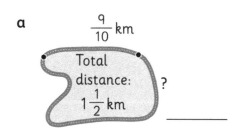

$\frac{9}{10}$ km

Total distance: $1\frac{1}{2}$ km ? _____

b

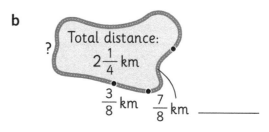

Total distance: $2\frac{1}{4}$ km

?

$\frac{3}{8}$ km $\frac{7}{8}$ km _____

c

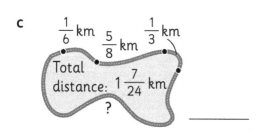

$\frac{1}{6}$ km $\frac{5}{8}$ km $\frac{1}{3}$ km

Total distance: $1\frac{7}{24}$ km

? _____

d

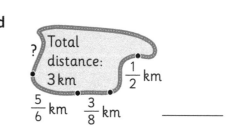

Total distance: 3 km

?

$\frac{1}{2}$ km

$\frac{5}{6}$ km $\frac{3}{8}$ km _____

Multiply Fractions

1 Complete the grid.

×	$\frac{1}{2}$	$\frac{1}{3}$	$\frac{3}{4}$	$\frac{2}{5}$	$\frac{3}{8}$	$\frac{4}{7}$	5
$\frac{2}{3}$							
$\frac{3}{4}$							
$\frac{5}{6}$							
$\frac{7}{8}$							
$\frac{9}{10}$							
$3\frac{1}{2}$							
$\frac{3}{5}$							
$\frac{7}{12}$							
$\frac{3}{2}$							

2 Work out:

a $\frac{1}{2}$ of 20 minutes _____

b $\frac{1}{10}$ of 4 m _____

c $\frac{1}{4}$ of 2 litres _____

d $\frac{9}{10}$ of $30.00 _____

e $\frac{3}{8}$ of $40.00 _____

f $\frac{1}{2}$ of 28 kg _____

g $\frac{1}{20}$ of 4000 m _____

h $\frac{2}{3}$ of $90.00 _____

Topic 21 Calculating with Decimals

Time Trials

Try to do each set of calculations mentally. Write the answers only. Record your time and your score.

Set A	Set B	Set C
0.2 + 0.6 =	1 – 0.2 =	0.8 + 0.8 =
0.3 + 0.6 =	1 – 0.8 =	1.2 – 0.4 =
0.9 + 0.1 =	2 – 1.5 =	1.4 + 2.4 =
0.5 + 0.6 =	0.8 – 0.7 =	3.5 – 2.7 =
1.2 + 1.2 =	4.5 – 2.3 =	2.3 + 4.8 =
2.4 + 4.5 =	1.2 – 0.8 =	5 – 3.7 =
0.8 + 1.2 =	1.9 – 0.6 =	7.2 + 6.2 =
1.3 + 1.5 =	2.3 – 2.1 =	2.4 – 0.6 =
4 + 0.9 =	1.8 – 0.9 =	3.9 – 0.8 =
2.8 + 0.5 =	2.3 – 1.9 =	4 + 2.7 =
0.9 + 0.7 =	4.5 – 0.8 =	3 – 2.7 =
2.9 + 2.8 =	3.1 – 0.5 =	2.55 – 1.2 =
3.1 + 4.5 =	3 – 1.5 =	0.99 – 0.55 =
1.1 + 3.9 =	9.2 – 0.7 =	2.45 – 0.4 =
2.7 + 3.3 =	7.8 – 3.4 =	3.2 – 0.8 =
Time:	Score: $\dfrac{}{15}$	Time:	Score: $\dfrac{}{15}$	Time:	Score: $\dfrac{}{15}$

More Adding

For each shape, write a calculation and work out the perimeter. All distances are in metres.

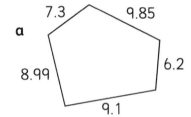

a

_____

_____

_____

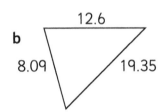

b

_____

_____

_____

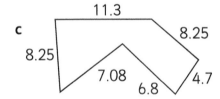

c

_____

_____

_____

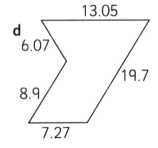

d

_____

_____

_____

Powers of 10

1 Fill in the operation symbol (× or ÷) and the power of ten used to get from one number to the next in each chain.

a

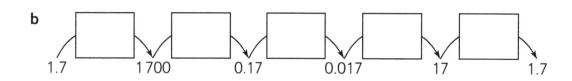

0.4 → 4 → 40 → 0.04 → 0.4 → 400

b

1.7 → 1 700 → 0.17 → 0.017 → 17 → 1.7

c

67 → 6.7 → 0.67 → 670 → 0.067 → 6.7

2 Fill in the missing numbers in each chain.

a

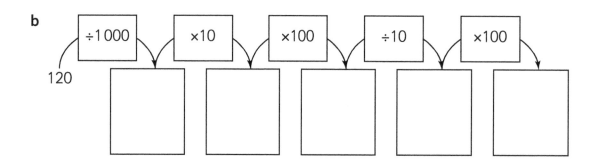

0.8 → ×10 → ×10 → ÷1 000 → ×10 → ÷100

b

120 → ÷1 000 → ×10 → ×100 → ÷10 → ×100

Multiply Decimals

1 Two digits have been accidentally erased in each calculation. Work out what they are and write them in the correct place.

a 2. ▨ × 8 = ▨ 0

b 12. ▨ × 10 = 12 ▨

c 7. ▨ × 3 = 2 ▨ .8

d 4. ▨ × 9 = 3 ▨ .8

e 3. ▨ × 9 = 32. ▨

f 9.2 × ▨ = 4 ▨

g ▨ .6 × 6 = 5 ▨ .6

h 0. ▨ × 8 = 4. ▨

i 4.5 × ▨ = 31. ▨

j 7. ▨ × 6 = 4 ▨ .6

k 5.7 × ▨ = 3 ▨ .2

l 9. ▨ × 3 = 28. ▨

2 Which set of bags is heavier? Circle it.

Set A
3.5 kg 3.5 kg 3.5 kg 3.5 kg 3.5 kg

Set B
4.1 kg 4.1 kg 4.1 kg 4.1 kg

3 Which set of ribbons will make the longer total length? Circle it.

Set A
6.8m 6.8m 6.8m 6.8m 6.8m

Set B
7.3m 7.3m 7.3m 7.3m

Divide Decimals

1 Divide.

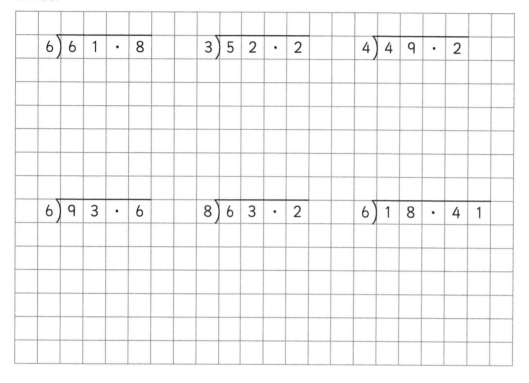

| 6) 6 1 · 8 | 3) 5 2 · 2 | 4) 4 9 · 2 |

| 6) 9 3 · 6 | 8) 6 3 · 2 | 6) 1 8 · 4 1 |

2 Complete the table. The division calculations in the two columns have the same quotients.

42 ÷ 0.2	gives the same value as	420 ÷ _____
57 ÷ 0.5	gives the same value as	_____ ÷ 5
3.28 ÷ 0.2	gives the same value as	_____ ÷ 2
85.2 ÷ 0.02	gives the same value as	_____ ÷ 2
0.27 ÷ 0.3	gives the same value as	_____ ÷ _____
0.625 ÷ 0.08	gives the same value as	_____ ÷ _____

Use a calculator to check your answers.

Topic 22 Perimeter and Area

Measuring Perimeter

1 Which units (mm, cm, m, km) would you use to measure the perimeter of each of these?

   a The coastline of Jamaica _____

   b A small triangle   _____

   c A desk in your classroom   _____

   d A fishpond in the garden   _____

2 Look at the following mobile phone covers.

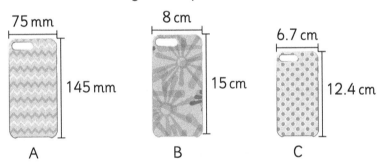

A   B   C

   a Which one has the greatest perimeter? _____

   b Which one has the smallest perimeter? _____

   c Calculate the perimeters and write them in order from the smallest to the biggest using the same unit of measure.

   A _____

   B _____

   C _____

Investigating Perimeter

Mr Johnson has a few goats that he wants to keep in a small fenced-off enclosure. He has 144 metres of fencing that he can use to make the enclosure. He would like it to have four sides and cover as big an area as possible so that the goats can move around and graze.

Work with a partner.

a Draw sketches and experiment with different shapes and sizes of enclosure. Each enclosure must have four sides and a perimeter of 144 m because that is the length of the fencing that Mr Johnson has.

b Use the table to keep a record of each enclosure you draw.

Shape	Measurements	Perimeter	Area	Comment
Rectangle	1 m wide 71 m long	144 m	71 m²	Not practical. Too long and thin; the goats will not have much space.

c Decide which shape and size is best. Show this to the rest of the class and tell them why you chose this shape.

Measuring Area

1 Draw the following shapes.

   a A square with an area of 16 cm².

   b A rectangle with an area of 12 cm².

   c Any shape with an area of 9 cm².

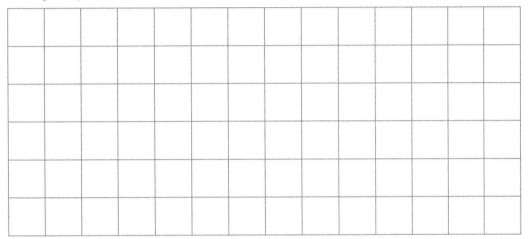

2 Complete this table. Calculate the perimeter and area of each of the rectangles.

Length (cm)	Width (cm)	Perimeter	Area
6	10		
2	5		
7	4		
3	8		
11	6		
2	7		
10	11		
6	4		
9	5		
12	8		

More Measuring

1 Work out the approximate areas of the following leaves. One square is 1 cm².

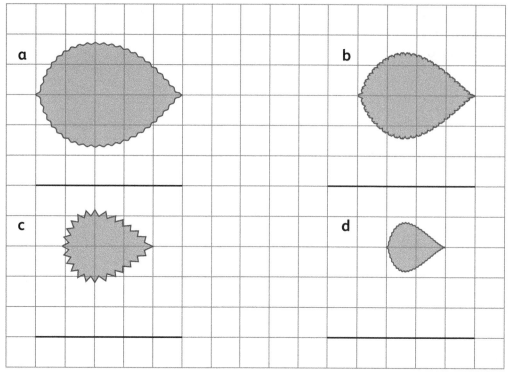

2 Work out the circumference and area of the following circles using formulae.

Radius	Diameter	Circumference	Area
3 cm			
4.5 cm			
22 mm			
	11 cm		
	8 m		

3 Draw two of the circles from the table above in your exercise book. Label the radius and the circumference of each circle.

Topic 23 Speed, Distance and Time

Speed as a Rate of Distance and Time

1 Convert the times **a–g** from minutes to hours. Use decimals in your answers.

   a 30 minutes = _____ hours

   b 45 minutes = _____ hours

   c 90 minutes = _____ hours

   d 15 minutes = _____ hours

   e 120 minutes = _____ hours

   f 40 minutes = _____ hours

   g 36 minutes = _____ hours

2 Complete the following table for three car journeys A, B and C.

Trip	Distance (km)	Time (h)	Speed (km/h)
A	325		100
B		0.166666666666	90
C	200	2.5	

How Fast Do You Walk?

To do this activity, you will need two stopwatches and a tape measure.

1 Work outside in an open space in groups of five.

    a Measure and mark out a distance of 50 metres.

    b Take turns to walk the distance as fast as you can while two other group members time how long it takes.

    c Record the results for each student on the table.

Student	Time Keeper A	Time Keeper B	Mean Time Taken	Speed in Metres per Second (Distance ÷ Time)

2 Work out the mean time for each student. Record this in the table.

3 Divide the distance by the time in seconds to work the speed of each student.

4 Rank the speeds in order from slowest to fastest.

_____, _____, _____, _____, _____

5 How could you improve the accuracy of the results in a race timed like this one?

_____

6 Linda walked for four hours at a mean speed of 5 km/hr. How far did she walk?

_____

7 Ramon took four hours to cycle 58 km on Sunday. Work out his mean speed.

_____

Topic 24 Probability

1 Colour in each spinner to create the probabilities given.

    a 3 in 4 probability of landing on red.

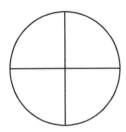

    b $\frac{1}{3}$ probability of landing on red.

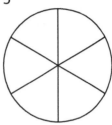

    c 3 in 8 probability of landing on green.

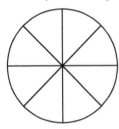

    d 40% probability of landing on yellow.

2 Explain in your own words what you understand by a fair game.

_____

_____

_____

3 Match each description to the probability. Write the letter.

    a Fairly likely   Zero   _____

    b Impossible $\frac{1}{2}$ _____

    c Highly likely   1   _____

    d Even chance $\frac{9}{10}$ _____

    e Possible but unlikely $\frac{7}{10}$ _____

    f Certain $\frac{1}{100}$ _____

Topic 25 More Measures

Different Ways of Measuring

1 Find out about the customary system of measurement. Then make a poster on this page to explain clearly how the measurements work and what the equivalents are in the metric system.

Standard Formats Recording Times and Dates

1 Fill in this application form, using the information in the notes on the side.

Application to join Mike's Sport Academy

Date: | Y | Y | Y | Y | M | M | D | D |

Notes about applicant

Name: Jon Millar

Date of birth: 16th February 2005

Date to join club: 2 September 2016

Phone number: 78929201

Name:

Date of birth: | Y | Y | Y | Y | M | M | D | D |

Date to join club: | Y | Y | Y | Y | M | M | D | D |

Phone number:

2 Write the following dates and times in ISO format.

a 10th October 1999

_____

b 2nd December 2016

_____

c 1st April 2000

_____

d 19th May 2006

_____

e He took 3 hours, 20 minutes and 5 seconds to complete the race.

_____

f 6 o'clock in the evening

_____

Hachette UK's policy is to use papers that are natural, renewable and recyclable products and made from wood grown in well-managed forests and other controlled sources. The logging and manufacturing processes are expected to conform to the environmental regulations of the country of origin.

Orders: please contact Hachette UK Distribution, Hely Hutchinson Centre, Milton Road, Didcot, Oxfordshire, OX11 7HH. Telephone: +44 (0)1235 827827. Email education@hachette.co.uk Lines are open from 9 a.m. to 5 p.m., Monday to Friday. You can also order through our website:www.hoddereducation.com

ISBN: 978 1 4718 6477 3

© Cloud Publishing Services 2017

First published in 2017 by
Hodder Education,
An Hachette UK Company
Carmelite House
50 Victoria Embankment
London EC4Y 0DZ

www.hoddereducation.com

Impression number   10 9 8 7 6 5

Year   2023

Cover photo © Thinkstock/iStockphoto/Getty Images

Illustrations by Peter Lubach and Aptara Inc.

Typeset in India by Aptara Inc.

Printed and bound by CPI Group (UK) Ltd, Croydon CR0 4YY

A catalogue record for this title is available from the British Library.